# SOLAR
# SYSTEM

# SOLAR SYSTEM

## DOROTHEA DePRISCO

**Interior Illustrations
by Robert Roper**

SCHOLASTIC INC.
New York Toronto London Auckland Sydney
Mexico City New Delhi Hong Kong

ISBN 0-439-20214-0

12 11 10 9 8 7 6 5 4 3 2 1                                        1 2 3 4 5 6/0

Printed in the U.S.A.                                                    01

First Scholastic printing, April 2001

vi

The words *solar system* refer to the Sun and all of the objects that *orbit*, or travel around, it. These objects include the nine planets, the Moon, the asteroid belt, comets, and meteoroids. The Sun is the center of our solar system. Mercury, Venus, Earth, and Mars are the planets closest to the Sun. They are called the inner planets and they are made mostly of rock. The outer planets are Jupiter, Saturn, Uranus, Neptune, and Pluto. These planets are large balls of gases with rings around them. Our solar system is part of a galaxy known as the Milky Way.

Scientists tell us that the solar system is about 4,600 billion years old. But how did it get started? They believe that the solar system was formed when there was an enormous explosion in space, which disturbed a cloud of gas and dust, called a *solar nebula*. This explosion made waves in space and the cloud began to collapse. As the cloud collapsed it grew hotter. Eventually, the cloud began to spin. It became really hot from spinning and grew more solid in the middle. Particles began to stick together and form larger groups of *matter*. Some of these groups got bigger, eventually forming planets and moons.

Today, most scientists think there are many other solar systems like ours. They have found nearly two dozen other planets surrounding stars that are very far away. It is incredible to think that there is more beyond our solar system!

Let this book be the beginning of your space travels. After you take a spin around our solar system, you can pick your favorite planet or star and keep reading!

- The order of the planets is Mercury, Venus, Earth, Mars, Jupiter, Saturn, Uranus, Neptune, and Pluto. How can you remember this? Memorize this sentence: My Very Eager Mother Just Served Us Nine Peas. Now take the first letter of each word and you have the first letter of each of the planets in their usual order!

HERE'S YOUR ASTRONOMY LESSON FOR TODAY, SON!

- Mercury is the closest planet to the Sun. There are only 36 million miles between

Mercury and the Sun. The second closest planet, Venus, is nearly twice as far away.

- It would be impossible to live on Mercury. Mercury has very little air, or *atmosphere*, surrounding it, so there aren't enough gases to reduce the amount of heat coming from the Sun. The temperatures can reach 750 degrees Fahrenheit during the day. That's five times as hot as the highest temperature ever recorded in the United States! At night, temperatures on Mercury can drop to as low as minus 320 degrees Fahrenheit!

- Mercury measures about 3,031 miles in *diameter*, which is kind of like the waist size of a planet, right around its middle. Mercury is just a little bigger than our Moon.

- Mercury moves around the Sun faster than any other planet. It travels about 30 miles per second, and goes around the Sun every eighty-eight Earth days. Compare that to the 365 days, or one year, it takes Earth to orbit the sun. Ancient

Greeks gave Mercury its name in honor of their swift messenger of the gods, Hermes. The Romans later translated the name to Mercury.

- If the Greeks had named this planet based on how long it takes to rotate on its *axis*, they might have named it the Slow Tortoise. Mercury completes a rotation about once every fifty-nine Earth days — slower than any other planet, except Venus. We have almost fifty-nine days before Mercury even has one day!

- Like Earth's Moon, Mercury is covered with craters. Many astronomers believe the craters were formed by meteorites or small comets crashing into the planet.

- Like Mars, Earth, and Venus, Mercury is made mostly of rock and metal.

- Mercury's brightness and size change as it orbits the Sun. Scientists call these changes phases. When Mercury appears from behind the Sun, it looks large and colorful. When it turns to the other side of the Sun it looks tiny and faint.

- In 1973, the *Mariner 10*, the only space probe (robot) to ever visit Mercury, was launched to examine the outside of this amazing planet. What did the robot find? Small deposits of water ice, hardly any atmosphere, but tons of rocks. Not exactly home, sweet home!

- The second closest planet to the Sun is Venus, a mere 67 million miles away.

- If all the planets wore belts, Venus and Earth could share the same size. They both have the same diameter: 7,521 miles.

- The temperature on Venus is a nice warm 867 degrees Fahrenheit. This is hot enough to melt lead!

WELCOME TO VENUS!
(READ THIS SIGN QUICKLY!)

- We can see Venus for several months each year, either in the morning or evening, with nothing but our eyes to help us. Venus is called both the morning star and the evening star.

- When visible, Venus is the second-brightest object in the sky, outshone only by the Moon.

- The air pressure on Venus would be enough to crush a human being, a building, or an entire city. The winds on Venus average 240 miles per hour. In comparison, the worst hurricanes on Earth can produce winds that are 155 miles per hour.

GOING TO VENUS ISN'T CHEAP! THE WIFE AND I JUST GOT BACK AND NOW WE'RE FLAT BROKE!

HOW YA DOIN'?

- Some spacecrafts have managed to land on Venus. They lasted less than an hour before the heat and air pressure melted and crushed them.

- The clouds on Venus contain sulfuric acid and cause rains that would burn and destroy anything that tried to live on the surface of the planet.

- Craters on Venus form in groups. This tells us that large meteoroids that make their way to Venus probably explode into pieces before reaching the surface and then land in bunches on the ground.

- If you haven't already guessed, Venus is the *hottest* planet. While Mercury is closer to the Sun, Venus is still hotter. Why? Because Venus has so much carbon dioxide in the atmosphere. Carbon dioxide traps heat and holds it in.

- Venus was a mystery until 1975, when the Soviet Union sent a space probe to investigate it. The robot took pictures of Venus that showed it had craters on the surface, just like Mercury and Mars but not as deep.

- Venus has a surface that has many mountains — some higher than Mount Everest — volcanoes, and plains filled with lava.

- Where do you think Venus got its name? Well, by now you know the Greeks and Romans named these planets thousands of years ago. Venus was named after the goddess of love and beauty because the planet is one of the brightest. Some call Venus the "jewel of the sky."

- The surface of Venus is covered with lava flows. The planet has several large volcanoes. Venus still has some eruptions, but only in a couple of hot spots. Over the past few hundred million years, Venus hasn't had any major eruptions.

- Astronomers refer to Venus as Earth's sister planet. Both are similar in size, mass, density, and volume. Both were formed at about the same time.

- Though Venus and Earth have similar characteristics, the differences are enough to keep you on Earth. There are no oceans on Venus and the planet is surrounded by a heavy atmosphere of carbon dioxide. The pressure is ninety-two times that of Earth's. If it were possible to walk on Venus, it would be harder than walking at the bottom of a swimming pool full of water. The clouds on Venus are made of sulfuric acid, an extremely dangerous substance.

- How old are you? How old do you think Earth is? It's 4.6 billion years old! But the oldest fossil found is only 3.9 billion years old.

- Earth rotates on its axis, turning one full turn every twenty-four hours.

- What is Earth made of? People, right? No, actually, it is 71 percent water. It also consists of nitrogen and oxygen.

- Below your feet is what is commonly known as the ground. Well, the scientific name for this is Earth's *crust*. Earth's crust is so deep that it extends even under the ocean!

- Below the crust is the *mantle*, which is even deeper and bigger than the crust. The mantle is important because it holds tons of metal-rich minerals for Earth.

- So, what's after the mantle? We're getting closer to the middle! Can you guess what scientists call the center of Earth? What do you call the center of an apple? Yes! The *core*!

- There are two cores inside Earth — the outer core and the inner core. The temperature in the inner core is 9,000 degrees Fahrenheit. That is hotter than the Sun's surface!

- The outer core is so hot that it is pure liquid. The inner core, which is also extremely hot, is solid because the pressure of Earth's weight holds the center together.

- When you look at the sky, doesn't it look like the Sun is close? Actually, it's 93 million miles away. Not exactly a quick trip, but Earth is the third planet from the Sun and closer than most of the other planets.

- One year on Earth is equal to 365¼ days. That means it takes Earth 365¼ days to orbit the Sun. Every four years we have an extra day, and that year we call a leap year. That is when February 29 is added to our calendar.

- Life on Earth could not exist without plants. Human beings would use up all of Earth's oxygen in 300 years. Plants are our buddies in breathing!

- Which celestial object is closest to Earth? Here's a hint: It has lots of craters and we send astronauts there! The Moon, of course! The Moon orbits Earth. When the Moon glows at night it only takes a second for its light to travel to Earth.

- The Sun's nearest known star neighbor is a small red star called Proxima Centauri. This star is 4.3 *light-years* away from the Sun.

- Why is our Earth so important? We need Earth for food, water, natural resources, and electricity. We need to take good care of our Earth. That means no littering or polluting and recycling our garbage!

YOU TAKE CARE OF ME AND I'LL TAKE CARE OF YOU!

- Mars is also called the Red Planet. With all of the minerals and iron in its soil, Mars has a rust color. But, it's not warm at all! It is cold and very dusty.

- Mars is the only planet whose surface can be seen in detail from Earth. This planet has many craters and canyons that give it a bumpy appearance.

- Mars is the fourth planet from the Sun and the seventh largest.

- In many ways Mars is similar to Earth. One day on Mars lasts twenty-four hours and thirty-seven minutes, while one day on Earth lasts twenty-three hours and fifty-six minutes! That means that Mars takes eighty-one minutes longer than Earth does to rotate completely on its axis.

- The surface conditions on Mars are more like Earth's than are those of any other planet. But the present life on Earth could not live on the surface of Mars. The surface temperature on Mars is much lower than that on Earth, rarely rising above the freezing point of water, which is 32 degrees Fahrenheit.

- Mars is the only planet other than Earth to produce evidence suggesting that it was once the home of living creatures. However, there is no evidence that life now exists on Mars.

EVIDENCE OF PREVIOUS
LIFE ON MARS

- Mars was named after the Roman god of war because the planet's red color reminded people of blood.

- What color do you think the sky is on Mars? Light blue? Nope, pink! Also, there are no clouds on Mars, only dust in the sky. The sky is pink when the temperature is minus 40 degrees. Then, when it gets even colder, ice clouds appear in a midnight-blue sky.

- Weather on Mars changes drastically — not like on Earth, where we slowly go into different seasons. Weather changes daily: one day a dust storm, the next day an ice storm. And the weather patterns are so wild that if they happened on Earth it would be like having an avalanche or flood every other day!

SHEESH!

TYPICAL MARS WEATHER

- The surface of Mars in the Northern Hemisphere (the top half of the planet) is smooth like Earth's. The Southern Hemisphere has craters and is bumpy just like the Moon.

- In the Southern Hemisphere of Mars, there is also a basin that is deep enough to swallow Mount Everest.

- There are grooves or impressions on the ground in Mars that show scientists that water used to flow on this planet. New information shows us there may be water there today, but since the weather goes from cold to super-cold on Mars, the water is in the form of ice.

- Mars has two moons. Their names are Phobos and Deimos. They are not round like other moons. In fact, they look like peanuts. Deimos is smaller, only ten miles long, while Phobos is seventeen miles long.

- Olympus Mons is the largest mountain in the solar system and is located on Mars. It is 78,000 feet high and has a cliff that reaches 20,000 feet.

- In 1984, a meteorite landed on Earth and because of its color and the chemicals that were in it, scientists believe it came from Mars.

- Get ready! NASA will be sending Space Rovers (machines that look like a cross between an army tank and a Range Rover) to Mars in 2003!

- Jupiter is the largest planet in the solar system. It would take over 1,000 Earths to fill up Jupiter!

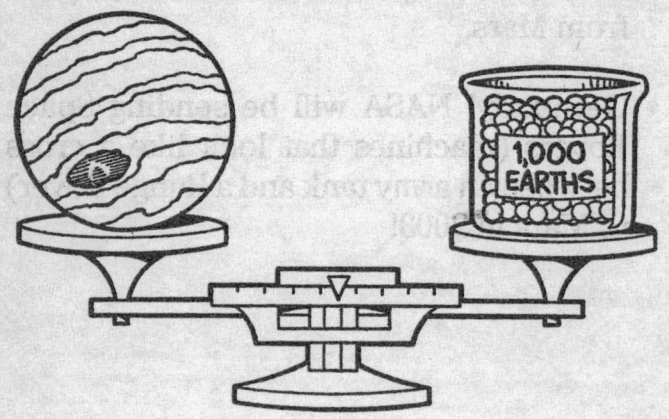

- Jupiter rotates faster than any other planet. It takes Jupiter nine hours and fifty-five minutes to rotate once on its axis, compared to the twenty-three hours and fifty-six minutes it takes for Earth to do so.

- Over 300 years ago, scientists discovered a giant red spot on the surface of Jupiter. This Great Red Spot is a really big storm.

This spot is also the coldest part of Jupiter. Scientists believe that this great storm stretches out so far that it equals the size of three Earths put together! The winds caught in this storm travel over 200 miles per hour.

- Saturn isn't the only planet with rings. Jupiter has three thin rings around its equator. However, they are much fainter than the rings of Saturn.

- Jupiter makes more energy inside itself than it gets from the Sun. In other words, Jupiter is so hot on the inside that it doesn't need the Sun as much as the other planets do.

- Jupiter is so hot that it is one of the brightest planets in the sky, as bright as a star!

- Jupiter is heavier than any other planet. Its *mass*, or quantity of matter, is 318 times greater than that of Earth.

- The force of gravity at the surface of Jupiter is up to 2.4 times stronger than on Earth.

A person who weighs 100 pounds on Earth would weigh as much as 240 pounds on Jupiter.

WOW! I'D BETTER LOSE SOME WEIGHT!

CRUNCH!

- Jupiter is also really strong and powerful. The planet has such a huge gravitational pull (that means it can pull in objects closer to it) that if a meteoroid floats by, it is going to get sucked toward Jupiter.

- Back in 1610, the Italian scientist and astronomer Galileo Galilei discovered that Jupiter had sixteen moons orbiting around it. He was particularly interested in the four large moons; that's why those moons are called the Galilean Moons.

- Saturn is the sixth planet from the Sun and the second largest in the entire solar system, with a diameter of 75,000 miles. Only Jupiter is larger.

- Saturn rotates on its axis faster than any other planet except Jupiter. Saturn completes one spin every ten hours and thirty-nine minutes, compared to twenty-three hours and fifty-six minutes for Earth.

- One year on Saturn equals twenty-nine and one-half Earth years.

- Saturn is the only planet less dense than water. What does that mean? If there were an ocean large enough, Saturn could float in it. Saturn is a very light and gassy planet, and it weighs very little!

- Watch out! The wind on Saturn blows at high speeds. Near Saturn's equator, the wind reaches 1,000 miles per hour!

- Saturn's rings make the planet one of the most beautiful objects in the solar system. Some people believe these rings came from larger moons broken by the impact of comets and meteoroids.

- The rings of Saturn are created as the planet moves. Gases, dust, ice water, and boulders spin around and create the rings. The major rings can measure 180,000 miles across.

- Each season on Saturn lasts about seven and one-half Earth years, because the planet takes about twenty-nine times as long to go around the Sun as Earth does.

7½ YEARS OF AUTUMN?! GET ME OFF THIS PLANET!

- Saturn is another giant gas planet. It is made up mostly of hydrogen and helium.

- Saturn can be seen from Earth without the help of a telescope, but its rings are not visible to the naked eye.

- Saturn was the farthest planet from Earth that the ancient astronomers knew about. The other planets were too far away for them to discover without the help of a telescope.

- Saturn has eighteen moons. That is more than any other planet. In 1995, scientists discovered that Saturn may have even more moons.

- What makes Uranus different from all the other planets? It spins on its side! Scientists believe that it spins on its side because of an accident with another planet millions of years ago.

- Uranus is the seventh planet from the Sun — 1.8 billion miles away. It was discovered in 1781 and was the first planet to be discovered with a telescope.

- This blue planet has a diameter of 32,190 miles, which makes it the third largest planet in our solar system.

- The atmosphere of Uranus is mostly hydrogen, with some helium and methane. Uranus seems to be blue-green in color because the methane gas of the atmosphere traps red light and does not allow that color to escape. Uranus may have red-colored bands like Jupiter's but they are hidden behind the methane layer.

- There has only been one expedition to Uranus. The U.S. *Voyager 2* flew by the planet in 1986.

- The *Voyager 2* spacecraft, at its closest, came within 50,600 miles of Uranus. The spacecraft needed special cameras to take photographs of the dimly lit planet. Uranus receives only $\frac{1}{400}$th of the sunlight that falls on Earth.

- It takes eighty-four Earth years for Uranus to orbit the Sun. You'll be a senior citizen before it gets all the way around! The length of a day on Uranus is seventeen hours and fourteen minutes.

HAPPY 1ST BIRTHDAY! -KEEP BLOWING, DEAR ...

BIRTHDAYS ON URANUS

- Neptune is the fourth largest planet and eighth from the Sun. However, sometimes Neptune becomes the ninth farthest planet from the Sun. During certain times, Pluto gets inside Neptune's orbital path and actually makes Neptune the planet farthest from the Sun.

- Neptune is 30,000 miles in diameter. It is four times bigger than Earth.

- Neptune looks like a big blue balloon. The element hydrogen gives off this color. Neptune is so far from us that when scientists look at it through a telescope it looks like a small blue star.

- Scientists can see the rings around Neptune with a telescope but unfortunately we cannot see them with just our eyes. Neptune also has seven moons that surround it.

- Triton is Neptune's largest *satellite*, or moon. Triton is about 1,680 miles in diam-

27

eter. Triton may once have been a large comet that traveled around the Sun. Scientists have discovered evidence that volcanoes on Triton once erupted with a smelly blend of ammonia and water. This mixture is now frozen on Triton's surface.

• The coldest temperature in our solar system is on Triton, which gives off a chilly temperature of minus 390 degrees Fahrenheit.

TRITON IS A LITTLE TOO COOL FOR MY TASTES!

- Neptune was the Roman god of the sea. What a perfect name for a planet that's blue like the ocean.

- Two astronomers discovered Neptune. John Adams predicted there was a planet near Uranus in 1843, but his boss didn't believe him. Then, in France, Urbain Leverrier started searching for a new planet. By 1846, it was agreed that both men should get credit for discovering Neptune.

- Neptune, Uranus, Saturn, and Jupiter are often called the gas giants. Made up mostly of gases such as hydrogen and helium, they rotate very quickly, usually have rings around them, are not very dense, and have many satellites or moons.

- One day on Neptune is equal to sixteen Earth hours.

- It takes poor old Neptune 164 years (in Earth time) to orbit the Sun!

- The Great Dark Spot on Neptune is really a tremendous storm brewing around the planet. There are winds surrounding the planet that were recorded at a speed of 1,200 miles per hour. This spot was discovered in 1989 when NASA sent *Voyager 2* (no astronauts, just the space probe) out to Neptune and Uranus.

- Neptune has eight moons; two of the largest are Triton and Nereid. Triton is bright, has a smooth surface, and spins in the opposite direction from Neptune.

- Pluto is not at all like the other planets. Pluto rotates very slowly, doesn't move in the same direction as the other planets do, and some scientists feel it isn't even a planet. Some think it should have been classified as an asteroid or comet. Some call it the accident planet.

HE'S DEFINITELY NOT FROM OUR SIDE OF THE GALAXY!

HAHAHA HAHAHAHA !!

- Pluto is mostly rock and ice and it is very small. Other planets have moons that are bigger than Pluto.

- Nobody knew about Pluto until the year 1930. It was so small and so far away, nobody could see it. Then American scientist Clyde Tombaugh, using a more powerful telescope, discovered Pluto.

- It takes small Pluto (it's only 1,800 miles in diameter) about 248 Earth years to make it around the Sun.

- Pluto has a moon — Charon. The moon is almost as large as Pluto, and it is so close to the planet that it shares its atmosphere. Many people refer to Pluto and Charon as double planets.

- Pluto is the smallest, coldest planet and usually the most distant planet from the Sun. Did you know that some of the moons in our solar system are larger than some of the planets? Jupiter's moon, Ganymede, the largest moon in the solar system, and Saturn's moon, Titan, are both larger than the planets Mercury and Pluto.

- We have never sent a spacecraft to Pluto because Pluto is so small and far away.

- A trip to Pluto may not occur until 2016. Study now and you may be the next astronaut to discover more about the solar system's most mysterious planet!

- Stars are big balls of glowing gas and dust in the sky. There are over 200 billion, billion stars in the universe. We can see only 3,000 without a telescope.

- Our Sun is the closest star to Earth. The next closest star is 25 million, million miles away.

- The Sun is the largest object in the solar system. If you measure the Sun all the way around its diameter, it would be 875,000 miles around. Its mass (weight) is 333,000 times larger than that of our Earth! The Sun is so big that you could take a million planets the size of Earth and fit them all inside it.

- The brightest object in the sky that we can see is the Sun, followed by the Moon, Venus, and Jupiter.

- In ancient times, people treated acne by looking for a falling star. They believed that if they found one, the star would wipe the pimples away as it fell.

I WISH I MAY, I WISH I MIGHT, GET RID OF ALL THESE ZITS TONITE!

- Even though our Sun is huge, it is not the largest star in the sky. There is an even bigger star called Betelgeuse.

- Betelgeuse is older than the Sun. As stars get older they get bigger, so it makes sense that Betelgeuse is bigger. Big Betelgeuse is about 8 billion years old and almost 500 times larger than the Sun.

- Our Sun is 4.5 billion years old, which is middle-aged — it is only halfway through its life!

- Stars are very hot. They can give off heat for 10 billion years.

- Want to hear something that's a little scary? Scientists believe that as the Sun gets older and larger, it will eat up the other planets before it becomes so large that it explodes. But that's not for another 5,000 million years or so.

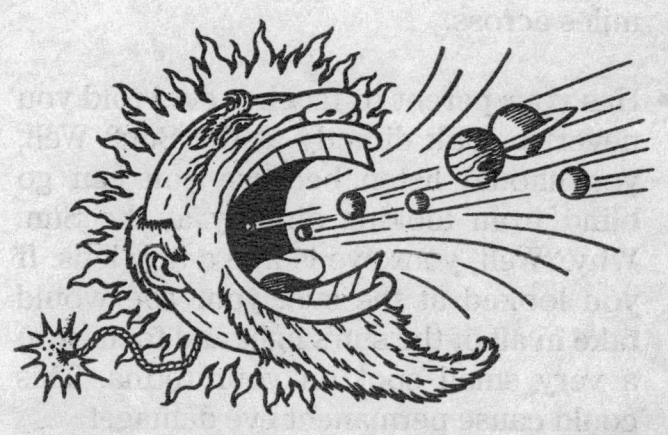

- Breaking news! A big, new star was discovered! And while it is not as large as Betelgeuse, it is brighter than Betelgeuse. Astronomers in Chile and Australia discovered the star, called R. Doradus. This huge red star is about 370 times larger than the Sun.

- Do you have freckles? So does the Sun! These "freckles" are small, dark patches called *sunspots*. Sunspots are parts of the Sun's surface that are cooler than other parts of the Sun. Sometimes sunspots cluster in groups and look bigger . . . sometimes as large as 7,000 million square miles across.

- Has your parent or teacher ever told you never to look directly at the Sun? Well, you should listen because you can go blind from looking directly at the Sun. Why? Well, your eye is like a tiny lens. If you looked at the sun, your eye would take in all of the sun's light and focus it to a very small spot on your retina. This could cause permanent eye damage!

- How long does it take for you to turn on the light in your bedroom? Just a second, right? Real light from the Sun takes eight minutes to reach Earth and about seven hours to reach the very end of the solar system.

- The Sun is a star, and like all the other stars in the universe, it was created from a *nebula*. What's a nebula? It's a large spinning cloud of gas. Everything inside that spinning cloud (tons of asteroids, meteors, comets, and even some moons) came together and formed the Sun.

- What is the most perfect star in the sky? The Sun, of course! The Sun is the perfect size, age, distance, and temperature to allow life to exist on Earth. If the Sun were smaller, younger, closer, or hotter than it is now, we might not be able to live on Earth.

- How long would it take to travel to the nearest star? Close to 167,000 years. To the nearest galaxy, far, far away? About 85 billion years. Here's some good news, though: It would take only eighty-six years to travel to Pluto, the most distant planet!

- Stars twinkle because starlight comes down to Earth through different layers of air, so the stars look bright, then dark.

- We know that the planets *orbit* the Sun. But what does the Sun orbit? The Sun orbits the Milky Way. It takes the Sun 225 million years to take a full spin around this galaxy.

- The Milky Way — no, not the chocolate bar — is a galaxy. Our solar system is in this galaxy. Galaxies are a collection of stars, gas, and dust.

- Do you know what kind of galaxy you live in? There is more than just one type! Some are shaped like ice-cream cones and we call them spiral galaxies. We live in a spiral galaxy.

THE HÄAGEN DAZ GALAXY

- A spiral galaxy has a bright middle and a couple of swooshes or arms that loop around the bright middle.

- An elliptical galaxy has a cluster of older red stars that have very little gas or dust. These galaxies can be in the form of a circle or oval shape.

- What do you think an irregular galaxy looks like? Really, it's just that — irregular. It does not have an identifiable shape — it looks like a bright cloud.

- The Milky Way has about 100,000 million stars. It is the band of light that is produced by the thousands of stars that lie in the main section of our spiral galaxy.

- You can see the Milky Way from your house. Just look up on a clear night. In the Northern Hemisphere, the best time to see the Milky Way is July to September and occasionally on a very clear winter night. In the Southern Hemisphere, try looking for it from October to December.

- But what does this amazing galaxy of ours look like? When you look up, you will see something bright that looks like a trail of snow or spilled milk. That's how it got its name . . . the *Milky* Way!

- We live in the very large and cool galaxy called the Milky Way, but do you know where the next galaxy is? The next galaxy is the *Andromeda* Galaxy, which is also a spiral galaxy that we can see with the naked eye.

- Did you know that there are about 300 stars with people's names? The names come from Arabic and were given to the stars 500 to 2,000 years ago. Today, astronomers name stars for their position in the sky, with just numbers and letters (their coordinates).

- What do you think it takes to be an astronomer, who studies the stars and space? Well, try studying calculus, physics, astrophysics, mechanics, electricity, and magnetism. That means college and graduate school.

- Inside galaxies, stars get together in clusters, sort of like groups of friends. There are Open Clusters, which are small groups of young, new, bright stars. The other type of cluster is Globular Clusters, which are much larger. The Globular Clusters are so large that they contain over 1,000 million stars per cluster.

- Our Moon is approximately 240,000 miles away from Earth. The Moon is about one-quarter the size of Earth. Earth's gravity causes the Moon to spin.

- Can you see the Moon on a clear night? Sure, you can! But does the Moon make its own light? No, actually it reflects the Sun's rays and that is why it looks so bright in our sky.

- The Sun's rays on the Moon cause the Moon to reach 254 degrees Fahrenheit! Ouch! At the same time, when the Sun's rays are not shining on the Moon, the Moon gets down to a very chilly minus 123 degrees Fahrenheit.

- Why do you think the Moon looks different to us each night? Well, as the Moon orbits Earth, it reflects light from the Sun. The sunlight illuminates different portions of the Moon's surface, creating what is known as the lunar phases.

- When you look at the Moon and see holes in it, what do you think they are? Have you ever said to someone that the Moon looks like Swiss cheese? Well, those holes are areas where lava has cooled down. Scientist call these *mares*.

SWISS CHEESE

THE MOON

THE SIMILARITIES ARE AMAZING!

- The Moon has mountains. One peak is actually as tall as Mount Everest. This peak belongs to the mountain range on the Moon called the Apennines.

- What happened on July 20, 1969? Neil Armstrong became the first person to step onto the surface of the Moon. He was on the *Apollo 11* mission.

- Since there is no wind on the Moon, the footprints left by the Apollo astronauts will remain there for many years.

- The Moon's gravity is one-sixth that of Earth; a man who weighs 180 pounds on Earth weighs only 30 pounds on the Moon.

- NASA (the National Aeronautics and Space Administration) was formed in 1958. If you want to become an astronaut, this is the agency you need to know about. NASA has sent space orbiters (probes, robots) into space since 1977.

- The latest space orbiter is the 1992 *Endeavor*, which cost $1.7 billion dollars.

- A space shuttle travels at 17,500 miles per hour.

- It takes 500,000 gallons of super-cold oxygen and hydrogen to fuel a space shuttle.

- When astronauts need to use the bathroom they do not have a toilet. They have a device that uses flowing air instead of water to move the waste through the system; solid wastes are stored on board the aircraft.

- Astronauts cannot shower on the space shuttle but they can take sponge baths.

- Did you know you need to be shorter than five feet eleven inches to be an astronaut? Limited cabin space in the shuttle prevents anyone taller from applying.

- The Hubble Space Telescope was launched in April 1990. This amazing device was named after the American astronomer Edwin Hubble. It can take pictures over a long period of time. For example, from the telescope data scientists can observe that Mars is becoming colder and drier.

- In its years of operation, the Hubble Space Telescope has observed approximately 8,000 objects.

- Used space shuttles are on display at NASA centers across the country. *Apollo 11* is at the Smithsonian Institution's National Air and Space Museum in Washington, D.C.

- Check out NASA's current missions at http://www.nasa.gov/

**Asteroids:** any of the small planets or minor planets.

**Atmosphere:** the mixture of gases that make up the air or climate that surrounds a planet.

**Axis:** an imaginary line around which a planet rotates.

**Crust:** the solid outer layer of Earth.

**Diameter:** the width of a circular object.

**Galaxy:** a huge collection of stars, gas, and dust.

**Light-years:** the distance that light travels in one year in a vacuum, or 5.88 trillion miles.

**Mantle:** the center part of a planet that is located between the crust and the core.

**Mass:** the quantity of matter that a body possesses.

**Matter:** the amount of material that occupies space.

**Meteor:** a meteoroid that is heated up when it enters Earth's atmosphere.

**Meteoroid:** a piece of rock tumbling through space.

**Nebula:** a cloud of dust and gas in space.

**Orbit:** the path of one object around another.

**Revolve:** to travel in a circle (or orbit).

**Rotate:** to turn around a center (or axis).

**Satellite:** a small object (such as a moon) orbiting a larger one.

**Solar system:** the Sun, together with the group of celestial bodies, such as planets, meteors, satellites, and comets, that revolve around it.

**Star:** a mass of gas that appears as a fixed point of light.

**Sunspot:** a dark, cool patch on the Sun's surface.

**Dorothea DePrisco** worked in publishing for seven years before moving to the island of Lana'i, in Hawaii. She taught high school English at Lana'i High and Elementary in Lana'i City. She currently lives in Los Angeles with her cats, Felix and Lucy, and works as a freelance writer.